# Earth Shine

*Anne Morrow Lindbergh*

# EARTH SHINE

*A Helen and Kurt Wolff Book*

*Harcourt, Brace & World, Inc., New York*

Grateful acknowledgment is made for the right to quote
From *Last Poems of Elinor Wylie*. Copyright, 1943, by Alfred A. Knopf, Inc. Reprinted by permission of the publisher.
From "Little Gidding" in *Four Quartets* by T. S. Eliot published by Harcourt, Brace & World, Inc. and reprinted with their permission.
From "Time's Cap-Poem" in *The Bright North* by Abbie Huston Evans, published by The Macmillan Company. Reprinted by permission of the author.

*Life* Magazine originally published earlier versions of both essays.

## Preface

Is there a bond between these two essays that they should be linked together in a book? Africa, a last stand of primitive life, and Cape Kennedy, summit of a scientific civilization—what have they in common? At first glance, nothing except their extremes. But today, even extremes are interrelated. Perhaps they always were, but we are now more aware of the inter-relatedness and the interdependence. Statesmen and businessmen, scientists and saints, poets and writers all tell us that the world is one. We cannot live in the wilderness and forget civilization. Nor can we live in the heat of civilization without realizing its increasing problems and without seeking answers or balances in the wilderness.

The sense of the earth as a whole, as a planet, is with us inescapably. We are unable to forget it or cut it in half in our minds. We think and feel not hemispherically but terrestrially, even though we do not always implement our vision. Teilhard de Chardin once said that he "wanted to express the psychology— the mixed feelings of pride, hope, disappointment, expectation—of the man who sees himself no longer as a Frenchman or a Chinaman but as a Terrestrial." And though there are few men who merit that title as well as he, there have been writers and poets, even before the days of flying, who have spoken as terrestrials. Henry Vaughan, the seventeenth-century poet, "saw eternity the other night/ Like a great ring of pure and endless light." In *Paradise Lost*, Milton envisaged Satan as he "winds with ease/ Through the pure marble air his oblique way/ Amongst innumerable stars, that shone/ Stars distant, but nigh-hand seem'd other worlds." Alone on a wintry hill at midnight, Thomas Hardy felt "the roll of the world eastward [as] almost a palpable movement."

With the expanded universe of Einstein, but before space travel, the planetary sense increased and found its voice in modern poets, like Abbie Huston Evans:

> *The terrible whorl of the Milky Way shines out*
> *To newt-eyes under; glory bears down ton-like;*

*Ordeal girdles us in. I marvel we live.*
*Yet live we do in the maelstrom, mites as we are;*
*On our acorn shook from the Oak, we ride out the dark.*

Fliers, even early airmen, not spacemen, have always had this sense. Looking down from above, they could discern more clearly the bones of earth and were aware, as Saint-Exupéry was, of how ephemeral is the flesh that clothes it—that fragile flower of life, growing on its surface, "like a little moss in the crevices."

But there are certain places on the globe where one has much more sense of the planet than in others: flying over the great deserts of Arizona and New Mexico; climbing for altitude over the endless glacial dome of the Greenland Ice Cap; and especially when the great continent of Africa first unrolls before the eyes, with its jungles and deserts, its serpentine rivers, its Great Rift, cracking across the baked earth, and its mountains rising majestically from the plains. Even to think of Africa is to see it as it is on a globe. That huge shoulder, bulging out into the Atlantic, is as unmistakable to earth men as it was to the astronauts who saw it from thousands of miles out in space.

Down on African ground, one becomes aware of the fecundity and variety of life that our planet nourishes.

Here, on the eastern highlands, are migrations of gnu and zebra inundating the plains of Serengeti, elephants browsing through trees in Tsavo, pink clouds of flamingo shimmering over Lake Manyara. And man, a Masai herdsman in his red robe, stands in the midst of his animals, as if he had been molded from the earth itself.

Nowhere else does one have such a view of the long slow evolution of life on earth: the ostrich who has outgrown his need to fly; the giraffe who has stretched his neck to reach leaves on the upper branches of trees; the swift gazelle, whose speed is perfectly matched to the speed of the cheetah; the colonies of baboons, banded together for survival; and even the foot-worn campsite of stone-age man who emerged from this fountain of life, and whose fossil remains have been unearthed in the gorge of Olduvai.

In Africa are the roots of life, the roots of man himself: the wilderness with which he struggled; the welter of life from which he rose; the animals he fought and over whom he won dominance. Here is no static Garden of Eden but a panorama of evolution as a moving process. One is aware of its direction; one looks deep into its past, and one wonders where it is bound.

I remember one night lying out on blankets next to our campfire in the high cool savannas near Mount Kilimanjaro. Flat on our backs under an endless roof of stars, we stared up into the Milky Way. It was a night when it was impossible to see the heavens as one did as a child: a round bowl dotted with silver-paper stars, some big and some little. In the vast spaces, we knew stars not as dim and bright, but as near and far. And we could distinguish dark holes in the fields of diamonds where there seemed to be no stars at all, where the eye drowned in inconceivable black depths of space. Looking into such a well of darkness, we were suddenly aware of a moving star. One of the myriad points of light was traveling. It was not the imperceptible change of position of a planet nor the swift blaze of a falling meteor; it was more like a jet plane strayed into a celestial field. "A satellite!" we said to each other in amazement. Perhaps Echo One or Two? In the wilderness, out of sight or hearing of other men, no light, no town, no highway visible, we were watching a man-made vehicle in space, one of the twentieth century's latest miracles. Toward this, we realized, man headed when he chipped that rough stone tool in Olduvai Gorge. The hand that first formed a tool had now made a rocket to leave the planet.

But if the inventions of modern civilization seem miraculous in the wilderness of Africa, the opposite is also true. In the canyons of a city's skyscrapers, in the smoking factories, in the whirring power plants of civilization, the mysteries of life, nature, and wilderness seem miraculous. From a spacecraft on its way to the moon, it is the earth which is the miracle, "a grand oasis in the big vastness of space."

The roots of life are in wilderness, even as the flowers of civilization rise from cities. But can one separate roots from flowers? Can the flower exist torn from its soil? We need wilderness as a source of life, but we cannot retreat from civilization. The Garden of Eden is behind us and there is no road back to innocence; we can only go forward. The journey we have started must be continued. With our blazing candle of curiosity, we must—like Psyche—make the full circle back to wholeness, if we are ever to find it. And it seems to be what we seek. The new studies of environment and the new exploration of space both bring us a terrestrial view. Glimpses have flashed upon us from the wilderness and from the skies. Now that we have seen ourselves from lunar space, the vision may be clearer, and the journey may mean more—

*We shall not cease from exploration*
*And the end of all our exploring*
*Will be to arrive where we started*
*And know the place for the first time.*

# Contents

# List of Illustrations

# THE HERON AND THE
# ASTRONAUT

# Cape Canaveral and Cape Kennedy

It was called Cape Canaveral when we first camped there with our children, over twenty years ago, behind the dunes in low palmettos and sprawling sea grape, only a few feet from the roar of the sea and the long white empty beach. Across the Indian River causeway was a sleepy Florida town, a row of stores, a hotel, a few well-kept palm trees flanking the post office, some outlying orange groves and stands of Australian pine, and leisurely traffic en route to more famous resorts, Miami or Palm Beach.

We had bumped over dusty one-lane roads to the wild deserted Cape, a sharp elbow in the Atlantic. Threading our way through stretches of scrub oak, slash pine, and palmetto, skirting around mangrove swamps, we found a camping site, in the lee of grass-covered dunes. A few paths, hot sandy spills worn by surf fishermen, cut through to the broad hard beach beyond, where heavy-winged pelicans sailed downwind in formation and schools of sandpipers danced in unison on the iridescent foam path of retreating waves. We found footprints of raccoon on the morning beach and on still days saw white herons, ankle-deep in the tideline. In the distance, a black-and-white-striped lighthouse stood guard over the long bare coastline.

This time we have come down to watch the launching of Apollo 8 for its moon-orbiting mission, scheduled for December 21, 1968. As our wheels touch the Florida landing field, we wonder if anything will be left of the wilderness we once knew. Nothing looks the same. Now the town has spread over broad highways, splashed with motels, ice-cream drive-ins, gas stations, sprawling supermarkets, and gaudily lit restaurants. The Cape itself, after one crosses the causeway, is now named Cape Kennedy, the great air and space center: "NASA." The roads are straight and paved, the jungle bulldozed back, and the grass on each side is neatly cut. The flat spit of land jutting out to sea is lined with rocket launching towers, curious skyscrapers of scaffolding, over thirty of them, standing like sentinels at intervals along the shore,

4

dwarfing our old striped lighthouse, which remains like a child's toy, relic from a forgotten past.

Since we are a day early for the launching, we are taken around the Cape to see the most striking features of the enormous NASA complex. The open-air museum is our first stop, where early rocket types are shown. In the stubbly grass we see these outgrown weapons of man's newest ventures in space: World War II rockets, a postwar Redstone, a Jupiter, an Atlas. Perched on their pads, pointing skyward, repainted to avoid salt air corrosion, they still look deadly enough to me; but in the fantastically swift growth of spacecraft, they are as dated as Civil War cannon in a country graveyard.

I feel rather like Rip van Winkle looking at this exhibit, for my last touch with spacecraft predates even these antiquities. In my early married life on a transcontinental flight, my husband and I stopped at Roswell, New Mexico, to see Robert Goddard, early rocket-pioneer-inventor. In his home town of Worcester, Massachusetts, the visionary physics professor had been dubbed "that moon man" by skeptical New Englanders, but on the dusty plains of New Mexico no one paid much attention to his converted windmill tower where the first unpredictable rockets twisted skyward. I never saw a launching, but I remember an evening sitting on a screened porch, while my husband and this quiet intense professor talked of space exploration. Flying was then a new adventure to me; I had just won my pilot's license, and these two men were talking of a step far beyond flying—ascent into space.

In half a lifetime, the age of space has come, I realize, as we move on from the dusty past of the outdoor museum to the most advanced "complex 39," where today's rockets are assembled. The mammoth Vehicle Assembly Building looms in the distance from the moment one enters the base. In the morning mist it seems to hang on the horizon like a big gray-and-white Cubist poster. As we approach, it takes on four dimensions, an enormous square mass surrounded by streets, smaller buildings, and parking lots filled with hundreds of pigmy cars. A city in itself—a city for giants. As one looks up at its monolithic sides one is given incredible figures of height and width. ("It's bigger than the Pentagon." "The lower bay is the size of the United Nations Building.") Inside, one is overwhelmed not only by size but by the tremendous complication. Four cavernous wells, or vertical hangars, yawn upward.

Here the rockets are assembled by giant cranes and tested by men who work on different parts from countless levels. When a rocket is finally assembled and checked, the building-high doors of the VAB draw open and it is moved out onto a "transporter," a huge platform mounted on four double tanklike "crawlers" for wheels. Erect on its transporter, with the mobile launcher alongside, it is slowly inched down the "crawler way" to the launching site where last preparations are made for flight.

This morning Apollo 8 is on its launching pad near the beach. We peer at it through the low fog but it is barely visible, obscured by the two service structures which enclose it like two halves of a shell.

Next to the mammoth VAB is a moderate-size building, about as large as a major air terminal, the Launch Control Center. There the telemetry is housed—the monitoring and control equipment. Through glass walls we can see rows of TV consoles with technical experts behind them watching and checking final prelaunch details, while another room is ready to observe the launching. Already on the wall a panel is flashing the countdown in diminishing minutes to tomorrow morning's take-off. As I watch, it winks disconcertingly from 20:23 to 20:22. I try to

7

relate this fantastically complicated organization to Goddard's tracking system, years ago, in New Mexico. The telemetry in Roswell, from behind a stout wall shelter, consisted of a pair of binoculars, an old alarm clock to drive a recording drum, and, of course, Esther Goddard's faithful movie camera. Esther Goddard was not only photographer, she was also secretary, and seamstress of parachutes in her husband's enterprise. I feel faintly nostalgic for the intensely personal aspect of the earlier era.

We are led to the Industrial Area of the Space Center, another city, where administrators, scientists, engineers, and technicians carry out the infinite steps of Apollo 8's prelaunch program. We see the Flight Crew Training Building where astronauts practice flight and landing procedures in Apollo and Lunar Module simulators. From the outside, these computerized machines are encumbered with boxlike incrustations, as if metal had gone berserk, growing shell on shell of armor. But inside they are exact replicas of the space vehicles, with complete instrument boards, operating controls and equipment, where astronauts practice maneuvers and adjust to artificial conditions seen in an artificial heaven.

I begin to be overwhelmed and rather oppressed by

the complexity, weight, and detail of what the astronauts lightly call the "hardware" of rocketry. And this complex, I realize, is only a latter stage of the rocket's development. There are no factories at the Space Center. Thousands of firms all over the United States have manufactured the myriad parts which make up the perfected vehicle. The giant Space Center we are being shown is only for final assembly, testing, and launching. I am staggered by the superhuman efforts that have gone into the program. One admires the meticulous care and precision of checking and training, but the layman feels crushed by the sheer weight and cold intricacy of this computerized, electronic, machine-oriented world. It would be unbearable if it were not infused with the genuine human sense of excitement and dedication of hundreds of teams of experts, and thousands of men, working together for a common end that, in this case, is not the killing of other men but the advancement of knowledge for all mankind.

With a sense of relief and surprise, we suddenly find ourselves at the door of the astronauts' quarters, staring at a sign saying that no one with a cold or symptoms of of a cold may pass beyond this point. Fortunately, there is not a sniffle in our group. The door opens to a human world. A small reception room is decorated with a Christmas tree, an artificial one, we are told, since a real one might have been a fire hazard. Tomorrow's astronauts, Colonel Borman, Captain Lovell, and Major Anders, with other astronauts are in an adjoining study. Grouped around a long table, they are looking at celestial maps and photographs of the moon's surface. A celestial globe stands at one end, and a large star map hangs on a wall. The astronauts are discussing with geologists what they should look for and may see on the moon. They greet us cordially and invite us to lunch—their last lunch on earth. We move into a dining room like a ship's mess hall. I feel as if I were on an ocean liner, isolated from the rest of the world. There are about fifteen men around the table, chiefly astronauts— not only tomorrow's, but yesterday's and the day after tomorrow's astronauts. On the wall are colored photographs of a Greek temple and a view from the White House showing the Washington Monument. The Apollo 8 looks almost as high as that to me. Imagine the Washington Monument blasting upward to the moon!

Lunch is hearty and very good; conversation is informal and relaxed. The astronauts and the aviator compare notes on wing-walking and space-walking.

The sensation of height, I hear, decreases with altitude until, to an astronaut space-walking in orbit, it hardly exists. There is no *up* or *down* in space, with earth simply *out there*.

My husband goes back forty years to his early encounter with space rockets and his first meetings with Robert Goddard. The year after his flight to Paris, he was already considering the next stage of human travel, and this led him into speculation about space. How could one overcome the limitations of wings and propellers? Rockets, jet propulsion, seemed the only answer. Couldn't rockets be used on airplanes to reach higher altitudes and speed, or even just to get emergency power in case of engine failure? The engineers and scientists he consulted were discouraging. Totally impractical, they pronounced flatly. A rocket burned fuel too rapidly. A combustion chamber lined with firebrick would be needed—too great a load for an airplane.

But several weeks later, a newspaper article caught his attention. The story described some experiments by a physics professor of Clark University. Dr. Goddard, according to the account, had launched a rocket in a field near Worcester, Massachusetts. The fiery blast-off, followed by the missile's erratic flight and crash

(Esther Goddard's parachute having failed to open), startled the neighbors, who thought a plane had fallen and exploded. As Dr. Goddard and his crew were salvaging the scattered wreckage, they heard a police siren and looked up to see a patrol car, two ambulances, and some alert reporters. Despite the professor's protests that he was conducting a series of safe and controlled experiments, the alarmed community and the Fire Marshal banned any further tests.

" 'They ain't done right by our Nell,' " was the disgruntled comment of a crew member, quoting the old barroom ballad. The rocket thereafter became known as "Nell"; but "Nell" could no longer operate in Massachusetts.

The lurid publicity about "the moon-rocket man" disgusted the scholarly professor, but it brought him a new supporter. My husband telephoned long-distance and drove up to Worcester the next morning. He spent the day talking to Goddard, looking at his designs, and seeing motion pictures of rocket tests. At last he had found someone who saw the possibilities of space flight. The rocket experiments must be continued in a more isolated environment, but where could backing be found for such a project? The Smithsonian Institution

and Clark University had already given some supportive grants to Goddard. My husband obtained an additional grant from the Carnegie Institute of Washington, but a much larger amount was needed. My husband turned to two men in the forefront of aviation development, Daniel and Harry Guggenheim. Already known for their faith in the new field of aeronautics, they had enough vision in 1930 to see the future in space, and to finance the development of rockets. Robert and Esther Goddard, with "Nell" and her dismantled tower, moved to Roswell, where the first major steps in the conquest of space were taken.

Goddard, my husband tells the astronauts, had ideas and dreams far outdistancing his designs. He had envisaged man's landing on the moon and even traveling to the planets, but he was cautious and practical when talking of the next step. Theoretically, he said, it would be quite possible to design a multi-stage rocket capable of reaching the moon. But, he had broken off, smiling at the idea of such a fantastic sum, it might cost a million dollars.

The group of astronauts bursts into laughter. It is a cheerful meal, with my husband shaking his head at the inconceivable amount of fuel consumed by an Apollo launching. In the first *second*, he figures out, the fuel burned is more than ten times as much as he had used flying his *Spirit of St. Louis* from New York to Paris.

The astronauts talk with frankness and detachment about the hazards of their flight to the moon; the absolute necessity of the engine functioning perfectly, the problems of navigation which must be accurate enough, on the return trip, to hit the crucial twenty-five-mile corridor back to earth. We have the impression of keen honest minds, sensitive perceptions, relaxed bodies, and eager spirits. "Think," one of them says almost boyishly, "it's hard to believe, this time tomorrow we'll be on our way to the moon."

If it seems unbelievable to them, how much more so to us. Yet we leave heartened and reassured from this encounter with the men at the center of this gigantic enterprise. Here in the midst of a scientific, mechanical, computerized beehive is the human element, the most exacting of all. There is nothing mechanical or robot-like about these men. Intelligent, courageous, and able, they inspire faith in human capabilities and in this particular human exploit of tomorrow morning.

# Night—The New Moon

**W**ould you like to go out and see the rocket tonight?" It is after dinner, almost midnight, and we
are to rise at four thirty to be on the Cape in time for
the launching. I hesitate a moment, tired from the
previous night's late plane trip from New York and the
exciting day's tour of the Space Center. "It's on the
pad. They'll be servicing it through the night—all lit up
with searchlights, quite a sight." Better than sleep, I
decide, as we pile in a car and set out. Already roads
approaching the Cape are crowded, the sides lined with
cars, tents, and trailers full of people spending the night
on the beach to be in place for the early morning
spectacle.

Even before we reach the Cape we see Apollo 8
miles away across the water, blazing like a star on the
horizon. We journey toward it until we are only a mile

or two distant. As we approach, it gets larger and brighter until it dominates the dark landscape, an incandescent tube, a giant torch with searchlights focused on it and beaming beyond over the heavens. The whole sky is arched with rainbows of light.

We climb out of the car and stand in the night wind, facing the source of light. Even at this distance we can see the rocket clearly, poised on its pad and gleaming white. The service structure, one half of its protective sheath, has been pulled away. Only the mobile launcher (the umbilical tower), that dark, bulky cranelike structure, stands beside it, dimmed by its brilliance.

For the first time the rocket is alone, whole and free. It is no longer in sections, dwarfed by the mammoth assembly building, or obscured by scaffolding. The thousands of details we witnessed this morning have been unified into a single shape. We cannot see the men still working on it from the swing arms of the launcher; or the pipes feeding it, the vital umbilical lines carrying fuel, electricity, air; or—except as a dazzling whiteness —the glaze of frost that coats it owing to the extreme cold of the liquid fuels. A wisp of vapor curls from one side like a white plume of breath in the darkness. All is simplified by distance and night into the sheer pure shape of flight, into beauty.

Seen from this dark field tonight, it has curiously biological overtones. The mobile launcher, that earth-bound structure with outstretched arms, looks like some giant insect. It is turned, almost seems to lean, with its overhanging crane, above the slightly smaller rocket, in a gesture that is half embrace and half release; while the rocket at its side is newborn, naked, silver-bright. Here is the seed split from its pod, the gleaming chrysalis cracked from its protective cocoon. Here is the new moon in the old moon's arms.

But, as with the newborn, there is something astonishingly tender about it. Vulnerable and untried, this is the child of a mechanical womb, of a scientific civilization—untried, but full of promise. Radiating light over the heavens, it seems to be the focus of the world, as the Star of Bethlehem once was on another December night centuries ago. But what does it promise? What new world? What hope for mortal men struggling on earth?

I think of Henry Adams at the turn of the century, standing, half awed and half repelled in front of the new powerful dynamo. Despite himself, he felt moved by "the inherited instinct . . . of man before silent and infinite force." In the end, he admitted, "one began to pray to it." Shaken by the experience, he tried to weigh the force of the Virgin who had built Chartres against the challenging force of the new power. The Virgin, he feared, "the greatest force the Western world ever felt," was being displaced by the dynamo. On this light-struck night, what is it that moves us—the worship of another dynamo?

And yet the symbols that spring to mind—even the words used by the experts—are inescapably human: "the umbilical tower," "the bird perched for flight," "the launch window is open." "The Spider." Can we escape from our humanness no matter where we go or what we do?

# Morning—"The Bird Perched for Flight"

We wake to the alarm at four thirty and leave our motel at five fifteen. The three astronauts must be already climbing into their seats at the top of their "thirty-six-story" rocket, poised for flight. The pilgrimage of sightseers has started to the Cape. Already the buses have left and lines of cars are on the roads. It is dark, a little chilly, with a sky full of stars. As we approach the Cape we see again the rocket and its launching tower from far off over the lagoon. It is still illumined with searchlights, but last night's vision has vanished. It is no longer tender or biological but simply a machine, the newest and most perfected creation of a scientific age—hard, weighty metal.

We watch the launching with some of the astronauts and their families, from a site near the Vehicle Assembly Building. Our cars are parked on a slight rise of

*19*

ground. People get out, walk about restlessly, set up cameras and adjust their binoculars. The launch pad is about three miles away, near the beach. We look across Florida marsh grass and palmettos. A cabbage palm stands up black against a shadowy sky, just left of the rocket and its launching tower. As dawn flushes the horizon, an egret rises and lazily glides across the flats between us and the pad. It is a still morning. Ducks call from nearby inlets. Vapor trails of a high-flying plane turn pink in an almost cloudless sky. Stars pale in the blue.

With the morning light, Apollo 8 and its launching tower become clearer, harder, and more defined. One can see the details of installation. The dark sections on the smooth sides of the rocket, marking its stages, cut up the single fluid line. Vapor steams furiously off its side. No longer stark and simple, this morning the rocket is complicated, mechanical, earth-bound. Too weighty for flight, one feels.

People stop talking, stand in front of their cars, and raise binoculars to their eyes. We peer nervously at the launch site and then at our wrist watches. Radio voices blare unnaturally loud from car windows. "Now only thirty minutes to launch time . . . fifteen minutes . . . six minutes . . . thirty seconds to go . . . twenty . . . T minus

fifteen . . . fourteen . . . thirteen . . . twelve . . . eleven
. . . ten . . . nine . . . Ignition!"

A jet of steam shoots from the pad below the rocket.
"Ahhhh!" The crowd gasps, almost in unison. Now
great flames spurt, leap, belch out across the horizon.
Clouds of smoke billow up on either side of the rocket,
completely hiding its base. From the midst of this
holocaust, the rocket begins to rise—slowly, as in a
dream, so slowly it seems to hang suspended on the
cloud of fire and smoke. It's impossible—it can't rise.
Yes, it rises, but heavily, as if the giant weight is pulled
by an invisible hand out of the atmosphere, like the lead
on a plumb line from the depths of the sea. Slowly it
rises and—because of our distance—silently, as in a
dream.

Suddenly the noise breaks, jumps across our three
separating miles—a shattering roar of explosions, a trip
hammer over one's head, under one's feet, through
one's body. The earth shakes; cars rattle; vibrations beat
in the chest. A roll of thunder, prolonged, prolonged,
prolonged.

I drop the binoculars and put my hands to my ears,
holding my head to keep it steady. My throat tightens
—am I going to cry?—my eyes are fixed on the rocket,
mesmerized by its slow ascent.

The foreground is now full of birds; a great flock of ducks, herons, small birds, rise pell-mell from the marshes at the noise. Fluttering in alarm and confusion, they scatter in all directions as if it were the end of the world. In the seconds I take to look at them, the rocket has left the tower.

It is up and away, a comet boring through the sky, no longer the vulnerable untried child, no longer the earth-bound machine, or the weight at the end of a line, but sheer terrifying force, blasting upward on its own titanic power.

It has gone miles into the sky. It is blurred by a cloud. No, it has made its own cloud—a huge vapor trail, which hides it. Out of the cloud something falls, cartwheeling down, smoking. "The first-stage cutoff," someone says. Where is the rocket itself?

There, above the cloud now, reappears the rocket, only a very bright star, diminishing every second. Soon out of sight, off to lunar space.

One looks earthward again. It is curiously still and empty. A cloud of brown smoke hangs motionless on the horizon. Its long shadow reaches us across the grass. The launch pad is empty. The abandoned launching tower is being sprayed with jets of water to cool it down. It steams in the bright morning air. Still dazed, people stumble into cars and start the slow, jammed trek back to town. The monotone of radio voices continues. One clings to this last thread of contact with something incredibly beautiful that has vanished.

"Where are they—where are they now?" In eleven minutes we get word. They are in earth orbit. They "look good" in the laconic space talk that comes down from over a hundred miles above earth. And one realizes again that it is the men above all that matter, the individuals who man the machine, give it heart, sight, speech, intelligence, and direction; and the men on earth who are backing them up, monitoring their every move, even to their heartbeats. This is not sheer power, it is power under control of man.

We drive slowly back to town. Above us the white vapor trail of the rocket is being scattered by wind into feathery shapes of heron's wings—the only mark in the sky of the morning's launching.

# *Afternoon—Merritt Island Refuge*

The long afternoon stretches before us. Instinctively we want to touch earth again, to drench ourselves in nature. The desire is as strong as sleep, or the thirst to plunge into the sea. We must find our human base, after the superhuman sight of this morning. Several astronauts have told us that the Cape is a wildlife refuge and, rather incredulous, we want to see what is left of the quiet coast we remember. The greatest rocket site on earth, with its noise, mechanization, buildings, and roads, hardly seems the place for a wildlife sanctuary. What could be further from primitive wilderness than this pinnacle of scientific civilization, this last reach of man, this rocket to leave the earth itself?

But actually NASA has preserved over fifty thousand acres of wild land, surrounding the central base of activity. The launching sites and Industrial Area are buffered from the mainland by a broad belt of wilderness. Here now, as before, migratory birds and water fowl, as well as birds and animals native to Florida, find sanctuary. The Cape, angling out into the Atlantic, has always been a crucial flyway for migrating birds. Without NASA's protection, it might well have gone the way of much of Florida's coastline, slashed by highways, cut up for real estate, and cluttered by road-side stands. Fortunately, it has been saved, not only for wildlife but also for man.

"Doesn't the blast-off frighten the birds?" I ask the refuge director, remembering the distracted flight this morning.

"It does," he answers, "but they come right back." There aren't enough launchings, he explains, to frighten them away permanently, as might happen near a commerical airport. "And also," he adds, hesitating, as though it might sound extravagant, "before launchings we try to go around with a helicopter and herd the wildlife to another area."

Out of sight of the rocket towers, we begin to realize, are wide stretches of jungle growth, fresh-water marshes and salt-water creeks, where great flocks of ducks and coots winter, and even occasional blue and snow geese find shelter. In the mangrove thickets, herons, ibis, and egrets hunt and nest. Hidden in the pinewoods and palmetto brush are white-tailed deer, bobcats, wild hogs, and opossums, as well as the raccoons we used to see.

Our pace eases as we drive over some of the same roads we covered on our tour of the Space Center yesterday, or even on our way to the launching this morning. Only a few hours ago, it seems days removed, another world. This afternoon we are looking for wildlife, not for launching towers and rockets. We have passed into a different frame of reference. The veil of civilization has become almost invisible. Instead of the mammoth gray-and-white VAB on the horizon, our eyes pick up white herons, standing in the ditches, poised motionless for prey. Instead of staring at the bulky simulators, we watch armadillos, small armor-plated animals, rooting back into the ages, nosing for insects in the turf. Instead of gazing at vapor trails, we sigh with delight as great blue herons lazily shake out their wings and glide to another bank. Like an umbrella, they unfurl their feathers for flight, and then abruptly telescope into a rigid walking stick to stalk for food.

Pushing to outer reaches of the Cape, we refind the kind of territory we camped on years ago. The same bumpy sand roads are there behind the rolling bulwark of dunes. On the beach beyond, pelicans still sail downwind and sandpipers eternally chase the shining edge of foam. Behind the dunes are saw-grass marshes where flocks of ducks beat the water white, taking off for flight, or, braking the air with their wings, make a great "whoosh" when they land, like a long sigh of contentment. Here a white-breasted osprey hides in the crook of a dead tree, waiting for the flicker of a fish in a pool below. Kingfishers dart across the streams like blue lightning.

Our eyes, quickened now to nature, find hidden surprises. That dark mass of sticks in a tall pine is a bald eagle's nest. That water-soaked log on a muddy bank is a sleeping alligator. His eye is glassy, but his crenellated tail twitches occasionally like a mechanical toy. Wild hogs scurry through palmettos and a red-tailed hawk spreads his wings as we pass. We stop to watch a water turkey, swimming noiselessly downstream, bowing its snaky black neck rhythmically with each stroke before it dives into the safety of the dark water below.

At sunset, refreshed by our deep plunge into the green silence of wilderness and the white stillness of beach and water, we turn back to the jangling

lights of town. In a faint pink sky above us, we see
the moon, a thin silver crescent with the shadow of the
full moon between its horns. "The new moon, with the
old moon in its arms," I think, remembering last
night's vision. There is a single bright star above it.
Not ours, though—not our star of this morning.
Invisible to our eyes, somewhere between us and that
moon, is Apollo 8 speeding ahead with its three
passengers perhaps seeing the same moon, or looking
back at us. The vast emptiness of celestial space
becomes alive because three mortal men are in it.
I remember how, in my early days of flying across
desert stretches of the West, the slightest touch of man
would light up a landscape. A dusty track or the round
ring of an abandoned corral caught my eye as a scrap
of glass on a road reflects sunlight. Here too, the dead
wastes of the universe have been touched by the spark
of life.

# Dialogue—Earth and Moon

In the days that followed, Christmas week, along
with the rest of the world, we watched through
television and radio the progress of the spacecraft on
its 500,000-mile journey around the moon and back
to earth. The voices of the control station at Houston
and the voices of the astronauts bounced back and
forth from spacecraft to earth, an incredible dialogue,
almost without interruption over six days and nights.
The terse technical space language, made up of astro-
nomical calculations, star sightings, and computerized
system checks—as vital as the heartbeats on a stetho-
scope—was interspersed with the very human com-
ments, impressions, questions, hopes, fears, and prayers
of three mortals in lunar space:

## CONTROL HOUSTON

You're on your way, you're really on your way now. . . .

. . . . .

## APOLLO 8

We see the earth now, almost as a disk. . . . We have
a beautiful view of Florida . . . we can see the Cape,
just the point . . . at the same time we can see
Africa. . . .

I'm looking out of my center window, which is
the round window, and the window is bigger than
the earth right now. . . .

The earth is very bright. . . .

It is a beautiful, beautiful view with predominantly
blue background and just huge covers of clouds. . . .
It is very, very beautiful. . . .

Waters are all sort of a royal blue, clouds, of course,
are bright white, the reflection off the earth is much
greater than the moon. The land areas are . . . sort
of dark brownish to light brown in texture. . . .

At a hundred and eighty thousand miles out in
space . . . what I keep imagining is if I am some lonely
traveler from another planet what would I think about
the earth at this altitude, whether I think it would be
inhabited or not. . . .

The earth is now passing through my window.
It's about as big as the end of my thumb. . . .

. . . . .

CONTROL HOUSTON

This is Houston . . . you are go for L.O.I. [lunar orbit insertion].

APOLLO 8

Okay. Apollo 8 is go. . . .

CONTROL HOUSTON

You are riding the best bird we can find. . . . All systems go. . . . Safe journey. . . .

We are standing by . . . we've got it, we've got it! Apollo 8 is now in lunar orbit. . . .

. . . . .

CONTROL HOUSTON

What does the old moon look like from sixty miles?

APOLLO 8

The moon is essentially gray, no color. Looks like plaster of Paris—or a grayish deep sand. . . .

The moon is a different thing to each one of us . . .

a vast, lonely, forbidding-type existence, great expanse of nothing . . . clouds and clouds of pumice stone— not . . . a very inviting place to live or work. . . .

The vast loneliness up here of the moon is awe-inspiring . . . it makes you realize just what you have back there on earth. The earth from here is a grand oasis in the big vastness of space. . . .

The sky is pitch black . . . the moon is quite light . . . a vastness of black and white, absolutely no color . . . forbidding, foreboding extents of blackness. . . .

■ ■ ■ ■ ■

APOLLO 8
(Christmas Eve)

We are now approaching the lunar sunrise, and for all the people back on earth, the crew of Apollo 8 has a message that we would like to send you:

> *In the beginning, God created the Heaven and Earth.*
> *And the earth was without form, and void; and darkness*
> *    was upon the face of the deep. . . .*
> *And God said, Let there be light. . . .*
> *And God said, Let there be a firmament in the midst*
> *    of the waters. . . .*

*And God said, Let the waters be gathered together*
*and the dry land appear. . . .*
*And God called the dry land Earth; . . . and God*
*saw that it was good.*

And from the crew of Apollo 8, we pause with good
night . . . and God bless all of you—all of you on the
good earth. . . .

. . . . .

CONTROL HOUSTON
We're coming up on that transearth injection maneu-
ver. . . . All systems are go, Apollo 8.


APOLLO 8
Roger, thank you. This is Apollo 8.

. . . . .

APOLLO 8
As we come back the earth looks pretty small from
here [97,000 miles].

. . . . .

APOLLO 8
How's the weather down there?

CONTROL HOUSTON

Beautiful. . . . They told us there is a beautiful moon
out there.

APOLLO 8

Now we were just saying that there's a beautiful earth
out there.

CONTROL HOUSTON

It depends on your point of view.

.   .   .   .   .

APOLLO 8

I think I must have the feeling that the travelers in the
old sailing ships used to have. Going on a very long
voyage away from home and now we're headed back
and I have the feeling of being proud of the trip, but
still happy to be going back home. . . . And that's . . .
richer than being right here. . . .

.   .   .   .   .

APOLLO 8

So, until then, I guess this is the Apollo 8 crew signing
off and we'll see you back on that good earth very
soon. . . .

Until finally, on the beginning of the seventh day, the spacecraft pierced through the narrow re-entry corridor, that tiny keyhole in space, into the earth's atmosphere, trailing a hundred-mile-long tail of light as it reached for home and splashdown, under blossoming parachutes, into a welcoming Pacific and a cheering world.

# Back to Earth

This overwhelming ovation, not only from the
United States but from the world—what does it mean?
The emotions aroused may still be too complex and
profound to reach the surface of words. Centuries may
pass before man can judge the significance of the Apollo
Program and the new perspective on man and his
planet—a view as revolutionary as the one Galileo illumined with his telescope.

Acclaim for the stupendous effort and achievement
of the program, the perfection of the flight, the gal-
lantry of the performers, is justified and understandable.
But does it explain such a wave of enthusiasm and
hope? The witnessing of sheer power is not enough
to cause our elation. The cataclysmic explosion of the
first atom bomb inspired feelings of awe, but not a

wave of hope; on the contrary, it engendered more of a sense of foreboding and fear. The supremacy of the machine alone—unmanned vehicles probing even farther into the universe—did not kindle such expectation.

The first step in the conquest of space, the material or scientific discoveries that will follow, the cracking open of a new electronic, computerized age—are these what mankind is cheering? The expanded horizons of science hardly explain the intensity of response. Deeper springs, one suspects, lie beneath our reaction.

What lifts our hearts today seems to be more in the realm of the human, the psychological, and the spiritual. Perhaps, as Konrad Lorenz suggests, space exploration safely absorbs man's aggressive and competitive instincts, and in applauding the astronauts' exploits, we are grasping at a hope of preserving peace on earth. Those noble qualities of man—heroism, self-sacrifice, dedication, comradeship in a common cause—which are tragically brought out in war, are evoked in many phases of the space development. And these qualities must continue to be aroused in some fashion, or man will cease to be all that man can be.

"Without adventure," as Whitehead wrote, "civilization is in full decay."

The homage to heroism, the challenge of adventure, the hope of a more peaceful world—all these have their part in our enthusiasm. But there seems to be another element in our response, a sense of recognition as well as discovery. A quality we had lost touch with, we dimly feel, has been refound. Some gap has been bridged, some conflict reconciled. Perhaps the unnatural rupture between man and the universe, which Malraux has ascribed to our machine-dominated civilization, was momentarily healed by the flight of Apollo 8, not alone by the feat but by the presence of the astronauts themselves. The cosmological revolution brought about by modern science was at last given a human face. Human eyes saw, human words came down; human gestures were watched. The Word of Einstein, Bohr, and Fermi was made flesh, and the world responded.

What one sensed through the astronauts was not simply the "triumph of the squares—the guys with the crew-cuts and slide rules who read the Bible and get things done." Nor was it "the worship of the dynamo," that arrogant belief that Western man, with his new scientific and technical powers, now has everything under control and can conquer the universe. What came through the experience, astonishingly enough, was a new-old sense of mystery and awe in the face of the vast marvels of the solar system, and a modesty before its laws. It was nearer to the reverence of a Thomas Huxley saying: "Sit down before fact as a little child," or the vision of a Teilhard asserting: "Our concept of God must be extended as the dimensions of our world are extended."

"Who is driving up there?" one of the astronauts' small sons asked his father in the spacecraft. The answer came down from celestial space: "I guess Isaac Newton is doing most of the driving right now"—a tribute to the law of gravity discovered by Isaac Newton, but even now not fully explained.

This refound harmony with the universe may be partly responsible for the surprising sense of release and renewal that followed the moon flights. We have been given another image of ourselves and our place in the cosmos. But it is an image that brings both pride and humility in equal measure. Pride in man's triumphs is balanced by a regained and vital sense of awe and mystery. "Given mystery," a poet has written, "we can endure." Today's mystery is not the old veiling by superstition of the things man does not understand, but a new unblinking gaze at the mysteries of the universe that may never be unveiled.

Another generation will judge what has changed, what is born, what is promised. We, who are here today, can witness only certain very close and tangible miracles that bloomed at this moment for men on earth. Because of the advance in science, mechanics, and electronics, man was able to achieve a giant step beyond himself into space—a step shared by all the world through the marvels of modern communication. And from this shared experience in the perceptions of remarkable men, another surprising and human gift came down to us. Along with a new sense of earth's smallness, a fragile, shining ball floating in space, we have a new sense of earth's richness and beauty, marbled with brown continents and blue seas and swathed in dazzling clouds—the only spot of color in a black and gray universe.

No one, it has been said, will ever look at the moon in the same way again. More significantly can one say that no one will ever look at the earth in the same way. Man had to free himself from earth to perceive both its diminutive place in a solar system and its inestimable value as a life-fostering planet. As earthmen, we may have taken another step into adulthood. We can see our parent earth with detachment, with tenderness, with some shame and pity, but at last also with love. As Elinor Wylie wrote of earth before man had circled it:

> It is not heaven: bitter seed
> Leavens its entrails with despair:
> It is a star where dragons breed:
> Devils have a footing there. . . .
>
> It balances on air, and spins
> Snared by strong transparent space;
> I forgive it all its sins;
> I kiss the scars upon its face.

With adult love comes responsibility. We begin to realize how utterly we are earth's children. Perhaps we can now accept our responsibility to earth, and our heritage from it, which we must protect if we are to survive.

Power over life must be balanced by reverence for life. For life, this rare and delicate essence, seems to be, as far as man's vision now extends, primarily the property of earth, and not simply life of man—life of animals, birds, butterflies, trees, flowers, crops. All life is linked. This is what makes up "the good earth."

As we left the beach at Cape Kennedy the last evening, our eyes followed a lone heron over the marsh, and rose with a cloud of wheeling duck on the horizon. We realized with a new humility, born of a new pride, that without the marsh there would be no heron; without the wilderness, forests, trees, fields, there would be no breath, no crops, no sustenance, no life, no brotherhood, and no peace on earth. No heron and no astronaut. The heron and the astronaut are linked in an indissoluble chain of life on earth.

Through the eyes of the astronauts, we have seen more clearly than ever before this precious earth essence that must be preserved. It might be given a new name borrowed from space language: "Earth shine."

# IMMERSION
# IN LIFE

Not long ago I spent a month on safari in the great animal preserves of East Africa, the high grasslands of Kenya and Tanzania. The word "safari" conjures up nineteenth-century pictures of British colonials in pith helmets with guns, filing through long grass, followed by lines of native bearers with equipment on their heads. Our family safari, organized by my husband, traveled by Land-Rover, carrying tents, bedding, food, and water.

Since my husband had been in Africa before with the men who know it best, friends from the Game Department and the National Parks, we took no guide. We used no bearers or guns. We went not to hunt but to see wild animals in their natural surroundings. And we found as romantic as any hunter's safari the big-game lands with names rolling out like drums: Serengeti, Manyara, Kilimanjaro, Kimana, Olduvai, Amboseli, Ngorongoro.

Coming back to gray skies and sober landscape in New England, I had a strong sense of deprivation. I felt curiously diminished, less alive. The reaction, I think, was not simply the shrunken horizons, nor the loss of flamboyant vegetation, trees and flowers. Much of East Africa is as desert-bare and brown as our Western plains. What I missed chiefly was the teeming atmosphere of life in which, for those weeks, we were immersed.

For to go to this part of Africa is to be immersed in another element as indescribably new as immersion in air—one's first plane flight; or immersion in water—skin-diving off coral reefs. For a brief period one seems to escape the limits of one's own species, the prison-bounds of a human body, as if one had shed a skin and become another creature with other senses and powers. What is this element? Isak Dinesen in her African stories describes it as air. But to me it is the intensity and variety of life itself.

We were plunged, day and night, in the life of wild animals who wandered at will and without fear before our eyes: gazelle cropping the grass in front of camp, zebra thundering past on the prairie, lions wandering across the road, giraffe nibbling at trees. Why should I, born and bred in New England, now miss these exotic animals I had never seen before? What have I in common with the mountainous elephant, the striped zebra, the towering giraffe, the swift gazelle? What did they mean to me?

Nothing, I would think, and yet I felt a connection with these animals I saw daily. There was some tenuous link between us, and now that it is broken I feel deprived, poorer.

I must explore this poverty, and this richness; for there must have been richness then, that I should feel poor now. What did I have for this month? What do I remember?

Flying to East Africa from Europe or America, one wakes to dazzling sunshine roaring in the plane window, almost an affront to eyesight. Below are great expanses of wild land stretching out in all directions: rolling plains, wooded hills, an occasional lake, a rim of distant mountains, and, very far away, one peak with a plume of snow. No concrete roads, no towns, no section lines. Occasionally the hand of man shows in a square of reforestation or in the glint of corrugated roofs. Then, suddenly, the green trees and white towers of the modern city of Nairobi.

One steps out at the airport into the dry heat of full summer. The dome of sky is pushed up overhead and the horizons are stretched out as far as sight can go. On all fronts the span is gigantic. One takes a deep breath and has the sense of a great continent.

My husband drives us immediately through Nairobi National Park, a tract of wild land near the airport and within sight of the tall buildings of Nairobi. The park is fenced on three sides but open to rolling hills on the south, so animals can come and go as they will. We drive over baked dusty roads, through dry plains of pale quivering grass and scrubby, stunted thornbush. We rattle down stony, serpentine valleys, where a hidden river is betrayed by a green trail of flat-topped acacia trees.

On the bare hills one begins to see unfamiliar silhouettes of animals against the sky. Giraffe first— slanting geometric lines, bending over bushes—like tipsy cranes on the horizon; the peaked humpbacks of gnu on a ridge; the high shoulders of hartebeest sloping down triangularly to hindquarters; the rounded buttocks of zebra; the delicately horned heads of gazelle, and feathery bushes that become ostrich when they raise long necks.

No other city in the world has such a zoo, where

animals are wild and unafraid, because protected. As long as you are in a car, they hardly move as you pass. Invisible, or at least disguised, in the Land-Rover, you are of no interest to them.

They are as close as if behind bars. But there are no bars; they are free—so free they seem tame, not because made subservient to man; tame, because innocent of fear.

We push through groups of zebra, cropping grass by the roadside, near enough to see the pinkish bloom of fine red hair that lies over their vivid black and white stripes. If you drive too close, they rear up on hind legs and wheel like plump circus ponies, all in a row. The thud of their hooves leaves behind a trail of golden dust.

Here are flocks of gazelle, with flashing white bellies and rumps, and faces patterned intricately as the corollas of flowers, sharp horns and ears pricking up like stamens and pistils. Looking up from their quiet cropping, they stand with hooves set precisely close, startled still for a second, like glass animals. Then, with the click of a car door, an explosion of hooves, heels, legs, and flicking tails, they are gone.

Farther on, in a grove of trees, are giraffe—soft in color and fawnlike, despite their great height and vivid spots. Their faces are gentle, almost human, with big wrinkled mouths and large surprised eyes under a coronet of ears and horny knobs. Peering down at you curiously from towering necks, often over a treetop as if trying to hide, they are rather prim, even prissy-looking—until they move. Then they are dream animals drifting across the landscape like milkweed down, or enormous insects floating long-legged over a meadow, feeding at flowers that are elm-sized trees.

There are colonies of baboons bouncing over a hill, and short-legged wart hogs trotting busily through stubble, heads overheavy with double tusks, whiplike tails erect as flags. And over a hill, in the shade of thorn trees, we come on a pride of lions sprawled in midday sleep. Three or four blond lionesses lie relaxed, poured out like honey in the sun. One of them rises to stretch—elbows to ground, haunches in air like a fireside cat. Another rolls over on her back playfully—a giant kitten, paws in the air. They look replete, drunk with sleep and sun, totally unaware of an audience.

The lion, sitting apart on a nearby slope, is aware but completely unconcerned. From time to time he turns his shaggy-maned head slowly in our direction, with neither fear nor interest. His flanks gleam in the sun; his head upraised, facing the wind, he sniffs and surveys the world. Tawny in the golden grass, he is totally right in his kingdom. As, in fact, every animal in the park seems to be in its right place. The only thing out of place is our gray fortress of a Land-Rover—and that glinting capsule in the sky overhead, the plane that has just dropped me in Africa.

But soon we are out on safari—in wilderness, far from Nairobi, airports, and civilization. We travel the hot dusty roads by day, putting up our tents in the cool of evening, sitting in front of the campfire at night, watching half-burned zebra-striped logs glow in the chilly darkness.

Africa, as one is always told, is a country of violent contrasts. Even the climatic contrasts within a single day in the highlands of Kenya or Tanzania are breathtaking. At midday the thin air trills with a dry heat. Bare hills burn in the copper dust. Quivering heat waves run like brush fire on the horizon. Eyes rest in relief on anything green—a line of flat-topped acacia trees floating like clouds on the airless savannas, or the four gentle peaks of the Ngong Hills. If one is out on the road, one counts the hours till sundown. In camp, shaded by trees, one faces the faint breeze and sits quite comfortably listening to continuous bird chatter. There are not only sounds one recognizes—monotonous tolling doves, whistling starlings, squawking hornbills —but a constant babble of unfamiliar noises: birds that sound like water bubbling from a bottle, birds that sound like tailors' scissors and mechanical toys, birds that endlessly repeat "hic-haec-hoc" or, more currently, just "ye-ye." One sits and waits for evening.

It comes swiftly, with the delicious balm of water after thirst. One is released from apathy, full of energy. The heat and dust of noon are forgotten. The air clears and is still; vistas expand between trees; one could advance unhindered to any horizon.

At our camp near Kimana in Kenya, we can see a distant line of ragged purple mountains to the north. To the south, Kilimanjaro's white top, cloud-covered all day, looms up enormous, gleaming, benign as the full moon. The savannas turn golden; acacia shadows lengthen to dark pools across the grass. Flocks of gazelle, cropping in the sun's late rays, are white as narcissus.

Night is again violent—cool, even cold on the high plateau. There is a sudden change of tension, a heightened quality of awareness. The mass of stars overhead beat down like rain. The profound mystery of darkness inundates the world.

A flood of sights, smells, noises, all alien to day, roll
in at dusk. The earth vibrates with sounds as the sky
vibrates with stars—crickets answering constellations.
Against a curtain of insect-drone, one hears the rumble
of lions, the eerie cry of hyena, sharp zebra barks,
braying gnu, shrieking baboons, and a thousand
unexplained hoots, snorts, and thuds.

One listens like a dog on the watch—as primitive
man once listened, in mystery and apprehension. It is
safe in the tent—a lantern hung on the ridgepole, the
fire flickering outside; and yet, one listens—one listens.

With predawn comes the benison of small-bird song,
innocent twittering in the trees overhead, announcing
to the world that night is over, that day has returned,
and those who have survived can greet the morning,
radiant and cool. The mountain reappears, newly
capped with snow. The prairie shimmers with a hint
of dew. Zebras move slowly through vistas of trees,
cropping the silver grass like animals of paradise.

But it is not paradise, nor the peaceable kingdom.
Lions made their kills last night, or at dawn; and now,
gorged, are asleep under trees. Hyenas have found the
carcass. Vultures circle overhead.

Another violent contrast; another turn of life's
wheel; another face of the wilderness to accept. For

Africa embraces both the bright and the dark, the benign and the cruel, the fleeting and the timeless, the swift and the ponderous—the impala and the elephant.

We see elephants in the evening on approaching our first campsite beside a dry riverbed. The pale dusty track angles in on our right, a luminous stream of sand pockmarked by animal hooves. Leaving our Land-Rover on a bank, we walk out onto the dry bed. We look both ways, and stop. Downriver stands a cliff of elephants—dark, enormous, motionless, their tree-sized legs rooted in sand, their ears widespread, great trunks facing us, white tusks gleaming in the dusk. They sway now, and blow dust from their sinuous trunks, huge ears flapping gently as palm fronds. They begin to move in single file along the bank, great shadows melting in and out of trees. In no hurry, they move deliberately, picking up one muffled foot after another with noiseless grace. We stand silently mid-river and watch the procession march in solemn rhythm, as if keeping time to inaudible drums.

Late in the evening, we hear them at the muddy pool across the river from our tent. They are pawing at the ground with ponderous feet to reach new water. They slosh in the mud and spray each other with their trunks. In the moonlight we can dimly see their hulking shapes;

and as we hear the water on their dusty hides, we feel ourselves a part of their joy.

Elephants, browsing through trees as lesser animals through bushes, give one not only another dimension of size and space, but another dimension of time. Their mammoth forms hark back to the world of mastodons. The slow, even rhythm of their march makes one feel they are moving to another time than man's. They have come from ages before us and are going somewhere we will never reach. Grizzled and wrinkled as old trees, they seem even older—old as hills or rocks, carved out of earth and imperishable as earth. The memory returns of the Indian myth about Brahma creating the first elephants, "the elephants of the directions of space," to support the universe.

I am haunted by time in Africa, time as history and time as rhythm, for they are allied, perhaps inseparable. The many strands of time constantly weave in and out of one's journey: time of the elephant and time of the impala, time of the European and time of the African— the many times of the many different Africans. For the more one travels, the further back in time one seems to fall. One comes by jet plane and arrives at a modern city, the capital of a new, vigorous African nation springing from the past of a British colony. Then, traveling over plains covered with wild game, one slips back a hundred years to the buffalo-filled prairies of our early West. Or, meeting great flocks of Masai cattle grazing dry hills, one slips further back to the Biblical days of Abraham and his descendants, counting their wealth in cows. Here a herdsman—some Masai David with his staff, erect in his earth-red robe—stands ready to defend his flock from lions.

Down inside the enormous extinct crater of Ngorongoro, guarded by its rim of mountain walls, one stands like Adam among animals fresh from the Creator's hand in a Garden of Eden. But near the Great Rift Valley, that giant geological fault cutting through East Africa, in Olduvai Gorge, where Mary and Louis Leakey discovered the skull of *Zinjanthropus*, one plummets down to stone-age man. Here are the rocks he chipped into tools, the bones he cracked for nourishment, and the round stones he set one above another in, perhaps, the earliest known dwelling made by man.

The rhinoceros, also, takes you back to another age. When you see his dark, mountainous back in the swamp, his heavy pendulous sides, his uplifted, probing horn silhouetted against the sky, you feel he has just

emerged from primeval slime. No words will describe him. You want a pointed stone to scratch on rock in crude sign language: "Look—what I saw today—like this—see!"

One is reduced to silence before rhinoceros, hippopotamus, and buffalo—not only dwarfed by their size, but speechless before their unfamiliar shape, stunned by their unmistakable power. One cannot say *ugly* or *beautiful*. By what standards to judge? One cannot say *astonishing* or *incredible*—for there they *are*. One can only say: Life—life here, too; life in this form, also.

We have our first glimpse of African buffalo in Kimana Sanctuary, where Game Warden Denis Zaphiro shows us—as he showed Hemingway years before—this idyllic small preserve under the shadow of Kilimanjaro. It is a hot afternoon and the Land-Rover jolts over a bare, volcanic plain. Each time we stop, a cloud of dust envelops us, hiding everything from view. Steering around black boulders, we approach a swamp. And there they are, black backs like great igneous rocks clumped together, heads down, grazing. They turn to look at us, heavy handle-bar horns curved on either side of black brows—ominous, lowering looks. Long, shaggy ears—drooping limply, curiously out of place—frame their suspicious faces. Then, as if on signal, they huddle together and move in unison, like a lava flow—all those black backs—into the marsh; into waving papyrus reeds, into tall grass and water cabbage, until they are covered up, engulfed by a sea of green. A few muffled puffs and snorts, then stillness again. Only the white egrets floating overhead show where they were. A smiling sea of grass hides all that tumultuous life.

Nowhere does one feel more strongly this sense of awe before an irresistible life force than in Serengeti National Park, Tanzania, watching the animal migrations. Across the vast plains, hordes of gnu, zebra, and gazelle move with the seasons in search of better grasslands. The first night zebra are all about our camp, barking continuously like neighbors' dogs. The next day, advised by park officials who constantly watch and study this seasonal trek, we go out to look for the army of gnu reported on the move. The endless plains, as far as we can see, are dotted with those gray, shaggy-maned animals, reminiscent of our buffalo. In the distance I think I see trees. No, not trees, I discover through my field glasses; forests of gnu stand on the rim of the world.

Seen in their migration, the gnu—cropping, shaking their hoary manes, starting up in a rocking gallop, fanning out always in the same direction, following some unfathomed law—seem to embody a primal force of nature. One stands in sober wonder before this tidal wave of life, overwhelmed by its power, but strangely revivified "by the sight of inexhaustible vigor, vast and titanic features."

And, inevitably, keeping pace with the gnu or zebra migrations, are the lion, the hyena, the jackal, trailing their game, picking up the weak, the young, the old, the unwary, for their sustenance.

By day the predators look peaceful enough—asleep under trees, replete and lazy. Lions in repose are not frightening. They look benign and bored. When disturbed by a car, they may rise with dignity and pace off, but not in deference to the disturber. They intended to move anyway. You are beneath their notice. When they turn their sultry, amber eyes on you, the light is alien and comes from a great distance, lit by fires from another world.

I am not much frightened even when I meet a pair of lions early one morning next to camp. We heard them rumbling around us all night, like distant thunder. I wake early; the sun is not yet up, but Kilimanjaro with its flat snowy cap fills the sky in front of me, as if it has just risen from the plains. The prairie, bounded by acacia trees, is soft with the milky light of morning. I walk cautiously toward a copse of trees and then stop, again hearing the unmistakable grunt of lions.

As if in a dream, a lion and lioness move out silently from behind the trees. Their backs are to me and they pace ahead into the empty prairie. I hold my breath, calculate swiftly that I am nearer to the camp than to them; wonder if they have seen me; and if they haven't, if they won't; and if they do, what will happen.

Then, at some breath of sound or scent, they turn, two heads in tandem, deliberately my way. No sense of alarm or urgency. They simply turn and look at me, the lion and the lioness. I make no motion; they make no motion. We stare at each other in silence for a long moment. Then, having satisfied themselves that I am neither dangerous nor worth eating, they turn away just as deliberately and continue pacing across the prairie. I feel honored that they have not considered me worth running away from—or toward. I have not disturbed the peace of the morning, and neither have they. The peace is unexpectedly disturbed when my son, coming out of his tent, catches sight of the scene. "Father," he calls out, more in surprise than alarm, "there are two

lions in the field and Mother is there with them."
At his voice, the lions abandon dignity and lope
away.

One must watch a lion in action, as at a kill, to be
frighteningly aware of its power. To find a kill, one
scans not the plains but the skies for birds of prey.
Some dark specks on the horizon, vultures circling
over a distant hill, arouse our curiosity. Keeping
an eye on them, we start off in their direction over
the trackless plains in the Land-Rover. The vortex
of birds rises and falls like a dust twister, but does not
move from the spot. As we approach a line of low
scrubby bushes, we find a water hole in a sluggish
stream. On the far bank a large collection of leaden-
winged vultures squabble among themselves, their
hooked beaks digging in under the striped skin of
a dead zebra. In their midst a spotted hyena tears at a
leg with its powerful jaws.

Suddenly the lioness appears from the bushes. She
lopes forward slowly, her stomach distended, already
gorged with her meal, perhaps aroused from sleep.
Action freezes for a second. Only her great rope of tail
twitches ominously at its singed black end. Then she
lunges—a whiplash of muscle, head shot forward, teeth
bared in a snarl—at the group around the carcass. The

hyena scuttles off. The vultures flap, rise, or hobble drunkenly back to a wider circle. The lioness is alone with her prey. Not really alone, for the hyena has retreated only a short distance, to an exact line acceptable to the lioness. As the zebra's killer, the lioness knows her rights. The hyena also knows the rules of the game and sits in his place like a reprimanded schoolboy. Fixed in his new position, he watches. The vultures, too, stand back in a huddle. They are not abandoning the site, only biding their time. On the far slope behind the kill, another audience, a line of zebras, watches from a distance, their silly heads all turned to look—alert, curious, but somehow unconcerned.

The lioness is aware of her animal audience but seems oblivious to our Land-Rover on the far side of the water hole. Her urgent problem is to guard her kill from the hyena and the vultures. She has already eaten, but there are other lions to feed and more meals in the carcass. How can she keep her prize? She begins to drag the dead zebra down toward the water hole—a difficult undertaking with a full-sized animal not half eaten. She can only manage to move it a few inches at a time. Picking up the zebra by the neck, she tugs, all her muscles taut. Then, letting go, she pants and turns her head, with its reddened jaws, to look at her competitors. They do not move—the hyena and the vultures. Still as statues, they stand their ground and watch. Their turn will come.

In the no man's land between the lioness and the vultures, a blacksmith's plover, immaculate in black and white plumage, bobs up and down in the grass, pecking for seeds or insects—as coolly detached from the drama as if by a pane of glass. And in the reedy pool between us and the lioness—where the unwary zebra must have stopped to drink—two little bronze-winged ducks move across the water, scarcely rippling its surface, innocent as birds in a tapestry.

Life and death, seen through the burning lens of Africa, are inextricably knotted. In the midst of such abundant life, death is absorbed and accepted; but never forgotten and hidden, as in more civilized worlds. Wherever one turns, it is there. The whitened bones of ancient kills dot a hillside like daisies. The blanched skull of a fallen impala, bare horns curved to an Apollo's lyre, rest as mute testimony to its fleet passage through life. The intense flame of life is matched by the sharp shadow of death. Flame and shadow are inseparable and of equal intensity. Life is not cheapened by the constant presence of death, but death is seen more clearly as an essential part of the pattern of life. Life in the lion cubs has been fed by the zebra who has eaten the grass. The fish eagle waits over the pool for fish and frogs who eat insects. The storks search the field for locusts which devour the crops. Each animal, following a law of survival in its feeding and killing, contributes to the ages-slow progress of evolving life.

We are just beginning to understand how necessary life is to other life, how delicate is the balance, what disaster disturbance brings—even to man. For we also are deeply involved in the balance of nature and dependent on its cycles.

Perhaps some of the tremendous renewal of energy one experiences in East Africa comes from being put back in one's place in the universe, as an animal alongside other animals—one of the many miracles of life on earth, not the only miracle. Religion traditionally filled this function by giving us a sense of reverence before the mysterious forces around us; but the impact of science on our civilization has created the illusion that there are no limits to our powers.

What a responsibility we carry, and what guilt when things go wrong—as they constantly do! No longer have we the faith to say, in the Quaker spirit:

"And whether or not it is clear to you, no doubt the universe is unfolding as it should." We are uneasy under the burden of our assumed omnipotence, like a child being forced into an adult role. Intuitively we know we are not gods, not omnipotent. "We need to witness," as Thoreau pointed out, "our own limits transgressed."

When the untamed forces of nature suddenly confront us it comes as a shock, but often a healthy one. It even brings a kind of relief, as all truth that one suspects and then discovers is a relief. And, like all truth, it illumines areas beyond the one on which it is focused. Areas we had forgotten, other sources of strength and security, come out of the shadows. In storms or blizzards, when the technological scaffold on which we depend fails, we rediscover the strong web of human relatedness. In the African wilderness, man refinds his ancient and eternal kinship with nature and the animals. He hears again the religious assurance: "You are a child of the universe, no less than the trees and the stars. [No less than the lion and the impala.] You have a right to be here." He regains some of the peace and beauty of old Chinese paintings, where the landscape towers above the tiny but vital figures of woodcutters or sages crossing a matchstick bridge or sitting in quiet confidence under a storm-racked pine.

The return to innocence is not, of course, possible except for brief periods. Even if we wanted to reverse the trend of civilization, we could not restore a virgin continent or the wild plains of our West. Nor can anyone stop the clock moving in Africa. The new African nations are going to grow, increase their population, their agriculture, their animal husbandry and industry. Inevitably, unless protected by parks and reserves, wild animals will lose their habitat, just as they have done in Europe and America. Wilderness is threatened everywhere. The extinction of animals is not the only danger; man faces the loss of a breathing space for all that is wild and free in his spirit. And not only his spirit, his physical welfare also, even his survival, is imperiled by the extermination of other life on this planet.

My mystical sense of connection with the animals in Africa may actually have a basis in reality. Animals are necessary to man, although man, insulated by his civilization, is often dulled to the need. Those newly returned from the wilderness are more aware of the deprivation; they feel starved for some element vital to their being. Coming back from Africa, I find my hunger eased by scattering corn for pheasants at the edge of my woods, or walking out over marshy eel-grass to watch wild ducks in the cove.

If the connection one senses with animals is an authentic bond, perhaps the renewal one feels in their presence has a deeper significance than we realize. Immersion in wilderness life, like immersion in the sea, may return civilized man to a basic element from which he sprang and with which he has now lost contact. Joined again to this primal current, he may find that "Life is as much a force in the universe as electricity or gravitational pull, and the presence of life sustains life."

Henry Beston felt this "sustaining force" on the dunes of Cape Cod. I felt it on the African plains. But one can find it anywhere one stops to look and listen. It can be cultivated in one's own country, one's own backyard. I have suburban neighbors who, with the remains of supper, coax raccoons to their hands; other who feed the wild geese, others who preserve the marshes as cover for pheasants, herons, and ducks. Even in cities one can find a touch of the wild by watching migrating birds in the parks. Whether a gazelle is nibbling the grass in front of your tent or a cardinal comes to your windowsill, the essential act is the same, here as in Africa.

In the wilderness the act is magnified to an over-whelming experience. One does not have to seek it out. It floods over one as naturally and gratuitously as sunshine, and like sunshine it renews. In such an atmosphere the connection between man and animal is brought to a burning focus. Concentration on the animal is total; distractions disappear.

It is not alone that the animals can exist in national parks and reserves, but that man, with his civiliza-tion temporarily withdrawn, can see. How often, in ordinary life, do we stop to look at a bird, an animal, a tree? In Africa one sees as if for the first time, like a child or an artist. Perhaps, even, like a saint, one focuses with that "absolutely unmixed attention" that Simone Weil equates with prayer.

And in the looking, something of a creative act takes place—a leap of imagination. One goes halfway to participation in the life of other creatures, feels a part of their actions. Something inside oneself leaps as they leap, or is quiet beside their cropping quiet, or watches with their alert watchfulness. One is stilled in their stillness—a stillness trembling with life, like the stillness of a flame. And in the moment of partici-pation, one has made the connection—or is suddenly aware of it. This act of imagination is an act of obeisance to life in another creature—life in an unfamiliar, yet related, form. And by this act, as by every act of imagination, one is enriched. For the act of obeisance to life, wherever one makes it, is in essence religious.